baby PRAIRIE DOGS

KIM THOMPSON

CREATIVE EDUCATION • CREATIVE PAPERBACKS

CONT

ENTS

I AM A PUP.

I am a baby prairie dog.

See my sharp claws? They can dig tunnels.
ear
fur
claws

I was born underground. Now I am six weeks old. I come out for the first time.

I play with my brothers and sisters. My mom feeds me her milk.

Our town is under a prairie. It has underground tunnels and rooms. It has about 20 families.

Outside, we watch for predators.

I find grass
to eat. I clip
it with my
sharp teeth.

I stay close to family and friends. We groom each other.

We touch noses to say hello.

SPEAK AND LISTEN

Can you speak like a pup?

Baby prairie dogs jump and yip.

Listen to these sounds:
https://www.youtube.com/watch?v=ghHCijUCrSA

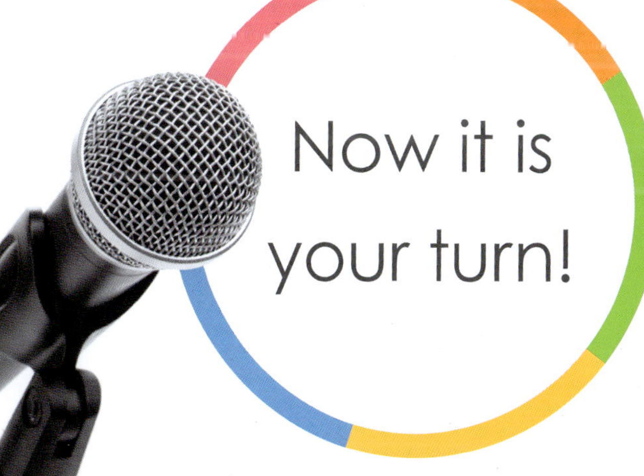

PUP WORDS

claws: sharp, hard nails on an animal's feet

groom: to take care of fur by combing it and picking out dirt or bugs

prairie: a large, flat area covered in grass

predators: animals that eat other animals for food, such as hawks and coyotes

READING CORNER

Christopher, Sophie. *Prairie Dogs in Their Ecosystems*. Minneapolis, Minn.: Bearport Publishing, 2024.

Patent, Dorothy Hinshaw. *At Home with the Prairie Dog: The Story of a Keystone Species*. Berkeley, Calif.: Web of Life Children's Books, 2023.

Schuh, Mari. *Prairie Dogs*. Mankato, Minn.: Capstone, 2020.

INDEX

PUBLISHED BY CREATIVE EDUCATION AND CREATIVE PAPERBACKS
P.O. Box 227, Mankato, Minnesota 56002
Creative Education and Creative Paperbacks are imprints of The Creative Company
www.thecreativecompany.us

LIBRARY OF CONGRESS CATALOGING-IN-PUBLICATION DATA
Names: Thompson, Kim, 1970- author.
Title: Baby prairie dogs / Kim Thompson.
Description: Mankato, Minnesota : Creative Education and Creative Paperbacks, [2026] | Series: Starting out | Includes bibliographical references and index. | Audience: Ages 4-7 | Audience: Grades K-1 |
Summary: "Introduce beginning readers to the world of baby prairie dogs with this life science starter. Includes photos, a labeled animal diagram, "Make a Noise" section, glossary, and further resources"-- Provided by publisher.
Identifiers: LCCN 2024043253 (print) | LCCN 2024043254 (ebook) | ISBN 9781640264229 (library binding) | ISBN 9781628329551 (paperback) | ISBN 9781640005860 (ebook)
Subjects: LCSH: Prairie dogs--Infancy--Juvenile literature.
Classification: LCC QL737.R68 T477 2026 (print) | LCC QL737.R68 (ebook) | DDC 599.36/7--dc23/eng/20250103
LC record available at https://lccn.loc.gov/2024043253
LC ebook record available at https://lccn.loc.gov/2024043254

DESIGN AND PRODUCTION
Design by Rhea Magaro
Production by Beeline Media and Design, Inc.
Art direction by Tom Morgan

PHOTOGRAPHS by Alamy Stock Photo/imageBROKER/R. Wittek, 5, Tierfotoagentur / M. Zindl, 8; Dreamstime/Alexandrebes, 10-11, Mikael Males, 7; Shutterstock/AllaR15, 12, Bohbeh, 13, Eric Isselee, 2-3, 4, JRDuncan, 9, Nick Fox, 11, pandpstock001, 14, Sabi Pollen, 6-7, Uttmurucha, cover

Printed in India